Mariana Pessôa Coelho
Elisabeth Regina Alves Cavalcanti Silva
José Gustavo da Silva Melo

Characterization of mangrove forests in the estuary of the Jaboatão River in the state of Pernambuco

Mariana Pessôa Coelho
Elisabeth Regina Alves Cavalcanti Silva
José Gustavo da Silva Melo

Characterization of mangrove forests in the estuary of the Jaboatão River in the state of Pernambuco

Structural analysis and environmental degradation

ScienciaScripts

Imprint

Any brand names and product names mentioned in this book are subject to trademark, brand or patent protection and are trademarks or registered trademarks of their respective holders. The use of brand names, product names, common names, trade names, product descriptions etc. even without a particular marking in this work is in no way to be construed to mean that such names may be regarded as unrestricted in respect of trademark and brand protection legislation and could thus be used by anyone.

Cover image: www.ingimage.com

This book is a translation from the original published under ISBN 978-613-9-67476-3.

Publisher:
Sciencia Scripts
is a trademark of
Dodo Books Indian Ocean Ltd. and OmniScriptum S.R.L publishing group

120 High Road, East Finchley, London, N2 9ED, United Kingdom
Str. Armeneasca 28/1, office 1, Chisinau MD-2012, Republic of Moldova, Europe
Printed at: see last page
ISBN: 978-620-8-16248-1

SUMMARY

ACKNOWLEDGMENTS

First of all, I thank God for the opportunity to be in this world and to live with such special people.

To my parents, who taught me so much in so little time together, or just enough time to keep going. Miss you.

To my family, who have been with me at all times, especially my grandmother, my aunt Valda, my uncle Meireles, my cousin Malu, Izabel and Pedro Lumumba, who have always believed in me and cheered me on.

To my university friends, especially: Clàudio Antônio Vieira, Carlos Vidal, Nelson Wilker (Juninho), Elizabeth Regina, Ana Paula Brandâo, Felipe André, Hewerton Alves, Patricia Passos, Tiago Henrique, José Gustavo, Neiva Marion and, of course, "Pseudo" Alexandre.

To UFPE, for the opportunity to broaden my knowledge.

To my advisor, Professor Maria Fernanda Abrantes Torres, for her patience, care and dedication.

To Fernanda Gomes Barbosa, who was fundamental to the development of this monograph. My sincere thanks to you.

And to Samuel Moraes, for his joy, companionship, trust and the courage to leave the comfort of his home to live this daily adventure that is living with someone. I love you, darling.

"What doesn't bring pleasure doesn't bring profit. In short, sir, study only what pleases you!"

William Shakespeare

SUMMARY

The mangrove is a tropical coastal ecosystem that colonizes sedimentary deposits formed by muddy, clayey or sandy vessels, occupying the intertidal zone up to the upper limit of the equinoctial high tides. It has a discontinuous distribution along the Brazilian coast, from Amapâ to Santa Catarina, and can present a *continuum* of distinct features depending on the profile of the coastline and the frequency and amplitude of the tides. In Jaboatao dos Guararapes/PE, there are around 5.7km^2 of mangrove vegetation, but these areas are heavily impacted by urban expansion, especially so-called irregular occupations. The aim of this study is to carry out a preliminary analysis of the structure of the mangrove forest on the right bank of the Jaboatao River, in the portion known as Ponta Cabo, located on the border between the municipalities of Jaboatao dos Guararapes-PE and Cabo de Santo Agostinho-PE, as well as to identify the main indicators of environmental impacts in the mangrove area. For structural characterization, the multiple plot methodology was adopted, with three plots being demarcated at a single site in the mangrove. *In* order to survey anthropogenic activities and their impacts, an *on-site* visit was made to apply a *checklist*. The mangrove vegetation at the site studied consisted of the species *Rhizophora mangle* L., *Laguncularia racemosa* (L.) C.F. Gaertn and *Avicennia schaueriana Stapf. & Leechman*, which showed a defined zonation pattern. *R. mangle* showed the highest density and relative dominance values, followed by *L. racemosa* and *A. schaueriana*. The structural analysis indicated a density of live trunks reaching 1,100ind/ha, with an average height ranging from 7.72m to 9m and a maximum basal area of 33.62 m2/ha, characterizing a mature forest. The level of environmental degradation in the sector analyzed was considered to be low (-90), with emissions of domestic and industrial effluents being considered the indicators of greatest environmental impact. This study shows that although the ecosystem is subject to discrete anthropogenic pressures, the site studied is within normal standards for a mature forest. It is important to note that the results obtained here could serve as a subsidy for future management and monitoring actions.

Key words: mangrove; structural analysis; environmental impact.

1 INTRODUCTION

The Brazilian coastline, a narrow coastal region made up of marine terraces, deltaic, estuarine and coastal plains, is around 7,408 km long and is home to most of the country's main economic, tourist and commercial centers. This region is home to important ecosystems, including estuarine ecosystems, which include lagoons, estuaries, river delta and mangroves (SCHAEFFER-NOVELLI, 2002).

The mangrove is a tropical coastal ecosystem that colonizes sedimentary deposits formed by muddy, clayey or sandy vessels, occupying the intertidal zone up to the upper limit of the equinoctial high tides. It has a discontinuous distribution along the Brazilian coast, from Amapà to Santa Catarina, and can present a *continuum of* distinct features depending on the profile of the coastline and the frequency and amplitude of the tides (SCHAEFFER-NOVELLI, 2002).

This halophilic environment of the intertidal zone is associated with typical vegetation cover, with the development of specialized flora, characterized by tree species that give it a peculiar physiognomy ("mangrove" feature).

This feature is exposed to daily washing by the tides, exporting particulate material (leaves, branches, propagules) to be decomposed in adjacent bodies of water (rivers, estuaries, coastal waters). In the innermost portions of the mangrove, characterized by relief that is sometimes higher ("apicum" feature) and sometimes by depressions, the litter ends up being decomposed on the spot, leading to the export of dissolved organic matter, which is of great ecological value for an area that goes beyond that of the ecosystem itself (SCHAEFFER-NOVELLI, 2002).

According to Walsh (1974), the best degree of mangrove development depends on five requirements: (1) tropical temperatures, with the average of the coldest month exceeding 20°C; (2) predominantly muddy substrates, consisting of silt and clay and a high organic matter content; (3) sheltered areas, free from the action of strong tides; (4) the presence of salt water, as mangrove plants are facultative halophytes and depend on this requirement to compete with glycophytes which cannot tolerate salinity; (5) a high tidal range.

The importance of these ecosystems goes beyond the ecological aspect; many communities living on the coast derive their livelihoods from the mangroves through artisanal and subsistence fishing; fishermen and shellfish gatherers derive their income from these activities. In this way, the degradation of this environment causes not only changes to the ecological environment, but also social and economic impacts.

In Brazil, mangroves are protected by the Forest Code under Law No. 4.771, of September 15, 1965, and recognized as Permanent Preservation Areas, but are threatened by various factors such as: the expansion of urban and port areas, predatory tourism and fishing, pollution from oil spills and domestic or industrial sewage, landfills and civil construction, timber extraction and disordered and illegal shrimp farming.

The environmental imbalances caused by these activities affect biodiversity and jeopardize the livelihoods of traditional populations who depend on these resources for their subsistence.

In the state of Pernambuco, the area occupied by mangroves is approximately 161.39km^2 in 2004, occurring in the lower reaches of large drainage basins and at the mouths of small coastal rivers along the coast (MONTEIRO et al., 2004).

According to Coelho et al. (2004) in Pernambuco, the mangrove forest extends from the mean tide level to the mean high tide level, between 1.0 and 2.0 m above mean sea level and 1.0 m above sea level on the terrestrial charts. The structure of the forest is mainly made up of the "red mangrove" *Rhizophora mangle* L. (Rhizophoracea), "siriùba mangrove" (*Avicennia schaueriana* Stapf. and Leechmam (Avicenniaceae), "white mangrove" *Laguncularia racemosa* (L.) Gaertn.f. (Combretaceae) and the "button mangrove" (*Conocarpus erectus*).

In Jaboatao dos Guararapes there is around 5.7km^2 of mangrove vegetation (ENGESAT, 2005), but these areas have been under a lot of pressure from urban expansion, especially from so-called irregular occupations. In addition, the municipality's estuarine areas are heavily degraded by industrial and urban pollution, landfills and garbage dumping (BRYON, 1994). The Jaboatâo River has been receiving loads of industrial effluents as a result of the expansion of the industrial park that has been set up along its watershed and domestic sewage from the urban centers established around it.

The aim of this study is to carry out a preliminary analysis of the structure of the mangrove forest on the right bank of the Jaboatâo River, in the area known as Ponta Cabo, located on the border between the municipalities of Jaboatâo dos Guararapes-PE and Cabo de Santo Agostinho-PE, as well as to identify the main indicators of environmental impacts in the mangrove area.

2 LITERATURE REVIEW

The structural characterization of mangrove vegetation is a valuable tool when it comes to the response of this ecosystem to existing environmental conditions, as well as studies and actions that lead to the conservation of the environment (SOARES, 1999).

Studies on the composition and structure of mangrove forests in Brazil have been carried out since the 1990s, including Soares (1999), Soares et al. (2003), Deus et al. (2003), Bernini and Rezende (2004), Menghini (2004), Silva; Bernini and Carmo (2005), Matni, Menezes and Mehlig (2006), Melo (2006) and Menghini (2008).

The vegetation structure and degree of disturbance of the mangroves of the Tijuca lagoon, Rio de Janeiro, were studied by Soares (1999), using the multiple plot methodology proposed by Schaeffer-Novelli and Cintrón (1986), and the temporal variation was also analyzed using aerial images. He identified a great variety in the structural development of the forest between the various sampling points, and this heterogeneity is considered a strong indicator of altered areas.

The structural and functional characterization of mangrove forests in Guanabara Bay, Rio de Janeiro, was carried out by Soares et al. (2003). The work presented results related to the action of tensors that determine different degrees of degradation and stages of regeneration (secondary succession) of the plots studied.

Deus et al. (2003) looked at the different histories of anthropization in the structure of the woody vegetation of three mangrove areas in Piaui. The vegetation was sampled using the multiple plot method, and the architecture of each phytocenosis was characterized using histograms of fixed 1m intervals.

The structure of the mangrove vegetation in the Paraiba do Sul river estuary was analyzed by Bernini and Rezende (2004), who used multiple plots distributed in the fringe and interior of the forest. This study found that the area analyzed showed better structural development of the forest when compared to other mangroves on the Rio de Janeiro coast, such as the Guanabara and Sepetiba bays.

Menghini (2004), in order to assess the degree of disturbance and the regenerative processes of the mangrove forests of Barnabé Island, Baixada Santista - SP, characterized the plant structure, leaf area, microtopography and litter production, and also observed leaf deformities and erosion.

The structural characteristics of mangrove forests in the estuary of the Sâo Mateus River, Espirito Santo, were analyzed by Silva; Bernini and Carmo (2005), using the multiple plot method. The study showed differences in the structural development of the forests, which correspond to variations in the frequency and periodicity of the subsidiary energies.

Matni, Menezes and Mehlig (2006) described the structure of three mangrove forests on the Bragança peninsula, in Parà, using the centered quadrant method (PCQM). The research concluded that the mangrove forests in this area are large, dominated by R. mangle. The study looked at the frequency of flooding as an important factor in the structural differentiation between the forests.

Melo (2006) identified the structure and zonation of the mangrove formations within the boundaries of the Guaraqueçaba Ecological Station - PR. This study found that the mangroves in this study area have well-developed forms from a phytosociological point of view.

Menghini (2008) dealt with the dynamics of natural recomposition in impacted mangrove forests on Barnabè Island (Baixada Santista) - SP. The results obtained refer to multitemporal analysis, natural recomposition, litter production, seasonality, microtopography and the characterization of forests at different successional stages.

As far as the state of Pernambuco is concerned, some studies have also assessed mangroves from a structural point of view: Souza and Sampaio (2000, 2001), in Suape; Schuler; Andrade and Santos (2000), in the Santa Cruz Canal; Correia (2002), in the Timbó river estuary; Nascimento Filho (2007), in the Ariquindà river estuary and Barbosa (2010), in the Pina mangrove swamp.

The mangroves of Suape - PE were researched by Souza and Sampaio (2000, 2001), who looked at the physiographic types and the temporal variation in the structure of mangrove forests, respectively. The results showed that, with regard to the distribution of species, there were no significant changes between 1988 and 1995 and no differences in the situations of anthropization. With regard to structural aspects, it was noted that eight years was not enough for the mangrove forest to recover.

Schuler; Andrade and Santos (2000) evaluated the composition and structure of the four species of mangrove forests found in the Santa Cruz Canal - PE. The authors characterized a greater abundance in the area of R. mangle, followed by L. racemosa and A. schaueriana, with the least frequent species being Conocarpus erectus. The study identified a 23.59% reduction in woodland between 1974 and 1988, while the urban area showed an increase of 625.03% over the same period.

The structural characteristics of the mangrove forest of the Timbó River estuary were analyzed by Correia (2002). The study evaluated the composition and structural characteristics of the mangrove using the plot method, and also assessed anthropogenic impacts through photographic records and the application of a checklist.

Nascimento Filho (2007) characterized the zonation patterns and structural development of the mangrove forests of the Ariquindà River, Tamandaré -PE, also considering aspects of the environmental gradient, degrees of flooding and salinity of the areas studied.

Finally, Barbosa (2010) carried out a structural analysis of the mangrove vegetation and a spatio-temporal diagnosis of the Pina mangrove, in which the author identified and quantified the spatio-temporal changes in the areas occupied by the mangrove over the course of three decades, as well as relating some environmental factors to the composition and structure of the forest.

Some studies have been carried out on the environmental analysis of the Barra das Jangadas estuarine system, formed by the joint mouth of the Jaboatâo and Pirapama rivers. According to Delgado Noriega (2005), due to the location and importance of this estuarine system for the coast, work has been carried out since the 60s. During this period, the following should be highlighted: Okuda and Nóbrega (1960), who determined the distribution and movement of chlorinity and the amount of flow; Okuda et al. (1960) researched the variation in pH, dissolved oxygen and the consumption of permanganate; Okuda et al. (1960) observed the variation of nitrogen and phosphate; Ottmann and Ottmann (1960) studied the sediments; Silva and Coelho (1960) carried out an ecological study in the estuary; Coelho (1963/64) researched the distribution of decapod crustaceans; Ottmann et al. (1965/6) studied the effects of pollution and the ecology of the estuary.

About thirty years later, in the 1990s, Coutinho (1997) analyzed marine erosion in the estuary and on the beaches of Piedade and Candeias; Silva (1997) related coastal dynamics to the meiofauna of an impacted environment in the Jaboatâo river estuary and Cunha et al. (1997) studied the morphodynamics of the estuary mouth and adjacent beaches in the state of Pernambuco.

At the beginning of this century some studies were published on the Barra das Jangadas estuarine system: Branco; Feitosa and Flores Monte (2002) related the seasonal and spatial variation of phytoplankton biomass with hydrological parameters; Noriega (2005) analyzed the spatial distribution of phytoplankton biomass and its relationship with nutrient salts; Noriega et al. (2005) studied the flow of dissolved inorganic nutrients.

More recently, some studies have looked at the Jaboatâo river estuary: CPRH (2006) presents relevant data in its Report on the Hydrographic Basins of Pernambuco; Barreto et al. (2006) characterized the land use and occupation of part of the estuary, between the municipalities of Cabo de Santo Agostinho and Jaboatâo dos Guararapes, using high-resolution images; GERCO-PE (2006) made a diagnosis of the physical and biological environment and a map of the land use and occupation of the metropolitan area of the Pernambuco coast, which includes this estuary; Hazin; Neto and Leite (2008) carried out an environmental diagnosis of the Jaboatâo and Pirapama rivers and Silva; Oliveira and Torres (2009) carried out a space-time analysis of urban development in the Jaboatâo river estuary, identifying the causes and consequences of the urbanization process.

From the above, it can be seen that most of the research carried out on the Jaboatâo river estuary is focused on hydrological parameters, with a lack of studies evaluating the distribution of vegetation, more specifically the occurrence of mangrove species, as well as the factors responsible for its degradation, which justifies this monographic study.

3 DESCRIPTION OF THE STUDY AREA

The Jaboatao river basin drains an area of 422 km2 and is home to a population of 446,426 inhabitants. It is made up of Precambrian rocks from the crystalline basement (90%), sedimentary rocks and sediments from the Tertiary-Qaternary cover (MOREIRA 2007).

The geomorphological units found in the basin include the crystalline modeling, the coastal tablelands and the coastal strip (MOREIRA 2007).

The most widely represented soil classes are indiscriminate mangrove soils, gleissols and quartzarenic neosols (ZAPE, 2002). According to Embrapa's new classification (2006), Gleissolos have the characteristics of a poorly developed soil with no B horizon. Gleissolo is a hydromorphic soil (saturated in water), rich in organic matter, which shows an intense loss of iron compounds.

The climate of this region is tropical hot and humid, type As', with autumn-winter rains, according to the Kopen classification (AYOADE, 2006), characterized by an average annual air temperature of 26°C and annual rainfall of over 2000mm. The area has two very distinct periods: the dry season, which runs from September to February, when the average monthly rainfall is less than 90mm, and the rainy season, which runs from March to August, with half of the annual rainfall occurring between April and June, where rainfall sometimes exceeds 350mm/month, according to data provided by the Department of Atmospheric Sciences at the Federal University of Campinas. Figure 1 shows the monthly averages for rainfall and temperature in the municipality of Jaboatâo dos Guararapes between 1911 and 1990.

Figure 1. Climatogram of Jaboatâo dos Guararapes/PE (1911 to 1990).

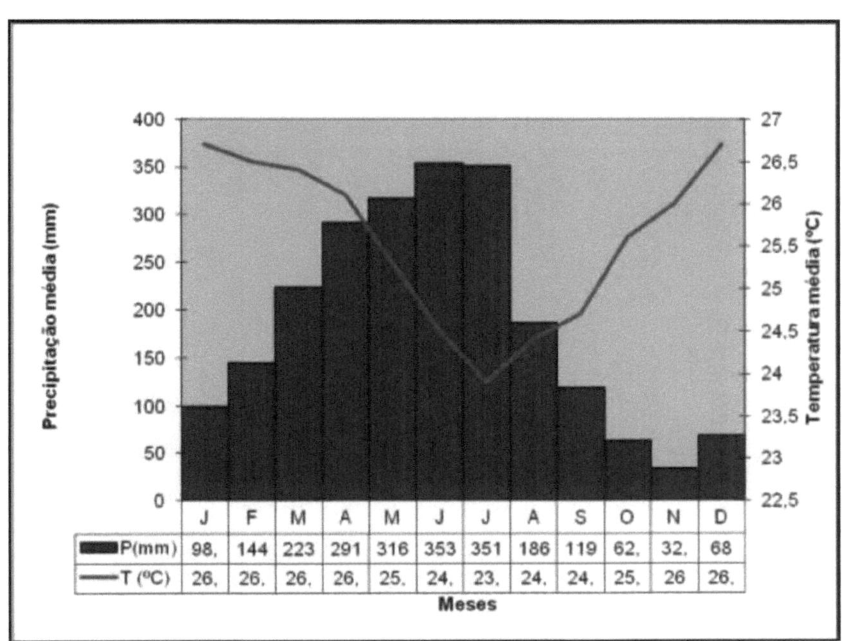

Source: DCA - UFCG (2018).

Land use is distributed between urban and industrial occupation, areas cultivated with sugar cane, polyculture and areas of forest and mangroves. The industrial activities installed in the basin include: chemical, food, metallurgical, textile, beverage, paper, plastics, electrical/communications, sugar and alcohol, clothing/artifacts/textiles, footwear, mechanical, pharmaceutical/veterinary and transportation (CPRH, 2006).

The Jaboatâo river estuary is located on the border between the municipalities of Cabo de Santo Agostinho-PE and Jaboatâo dos Guararapes-PE, at coordinates 8°13'/ 8°16' S and 34°58'30"/ 34°56'30" W, near Pontezinha and Ponte dos Carvalhos, at a distance of 20km from the city of Recife. It is shaped like an elongated "S", shallow and small, with a width of between 200m and 250m and a straight-line length of approximately 3km. It flows into the Pirapama River, forming the Barra das Jangadas estuarine system, which together drains an area of around 1,000 km2 up to the mouth of the Atlantic Ocean (NORIEGA et al., 2005) (Figure 2).

Figure 2 - Spatial location of the Barra das Jangadas/PE estuarine system.

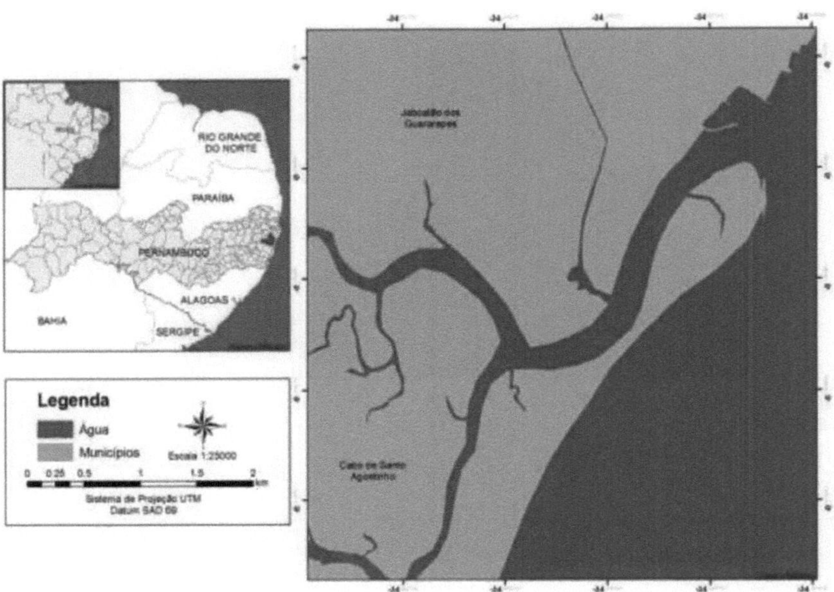

Source: Prepared by the authors (2018).

The urban expansion of the municipality of Jaboatâo dos Guararapes took place mainly from 1940 onwards, when, according to Hazin; Neto and Leite (2008), the municipality began to show an accelerated process of demographic densification, with a growth rate similar to that of the municipality of Recife.

This occupation has reduced the mangroves in the areas close to the Jaboatâo river estuary; these areas have been deforested and landfilled for the construction of housing developments and allotments. The consequence of this occupation process is strong demographic pressure on the mangrove ecosystem, reducing the vegetation and modifying the natural tidal flows, altering the entire life cycle on the edges of the mangrove areas (HAZIN; NETO; LEITE 2008).

Finally, still in relation to the characterization of use and occupation, Barreto et al. (2006) found that the estuary of the Jaboatâo River has two completely antagonistic situations: on the left bank, where anthropic action puts the maintenance of natural environmental protection structures at risk, and on the other bank, the exuberance of the limestone sandbank is preserved in its splendor.

4 MATERIAL AND METHODS

1.1 Structural Parameters

For the structural characterization of the forest, a site was delimited on the right bank of the Jaboatâo River, in the portion known as Ponta Cabo, on the border between the municipalities of Jaboatâo dos Guararapes-PE and Cabo de Santo Agostinho-PE, with the coordinates of 8°14'27"S and 34°56'41"W as its midpoint (Figure 3).

Three 10x10m plots (A, B and C) were analyzed at the site, oriented perpendicular to the river bank, following the physical-chemical gradient. The analysis was carried out on November 3, 2009, at low tide of 0.15m at 09:45 h.

Figure 3 - Spatial location of the site in the mangrove swamp of the Jaboatâo River, PE.

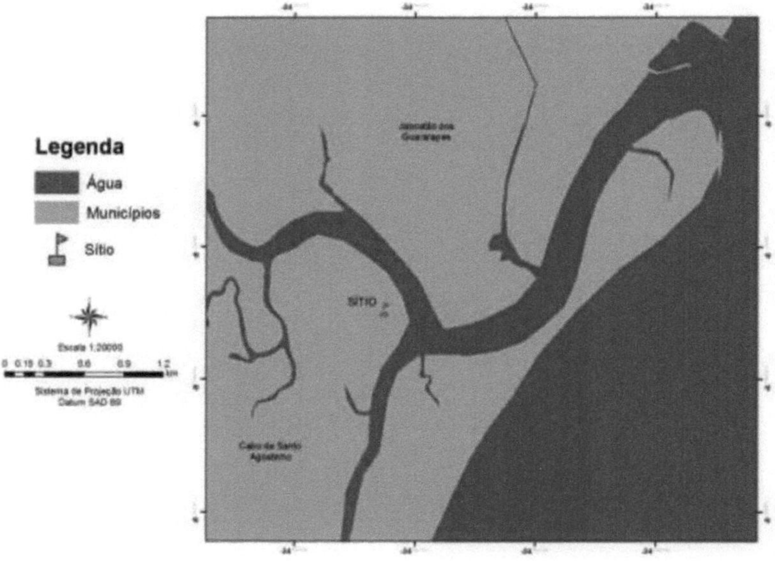

Source: Prepared by the authors (2018).

In all the plots, the soil temperature was measured with a spit thermometer and the air temperature was measured with a graduated Incoterm thermometer, as well as the salinity of the interstitial water at a depth of 30 cm. This last parameter was measured using an Instrutherm optical refractometer, scale 0 to 100.

The structure of the vegetation was analysed according to the methodology described in Cintrón and Schaeffer-Novelli (1981) and Schaeffer-Novelli and Cintrón (1986), which consists of using multiple plots to characterize the structural and functional properties of the system.

To apply the structural parameters, some characteristics of the forest were taken into account, such as: species composition, diameter at breast height of live trunks (DBH), basal area of live trunks (BA), height of individuals, number of trunks, density of individuals per hectare, dominance and frequency.

Nylon rope and a calibrated measuring tape were used to mark out the plots. As each tree was being catalogued, it was marked one by one with string to prevent the individuals from being recounted.

The methodologies used to measure these structural variables are described below:

1.1.1 Tree Diameter (DBH)

The diameter of the trees was measured using rules that obeyed the irregularities of each one. We used measuring rods graduated in centimeters and the values were divided by $\pi = 3.1416$ to obtain the diameter of each individual. The DBH was measured at a height of 1.30m from the ground using a measuring stick. The diameter measurements were grouped into classes of trunks with diameters <2.5 cm, ≥2.5 cm and ≥ 10 cm.

The average diameter is defined as the diameter of the tree with the average basal area. This value allows comparison between forests and can be correlated with other structural measures. The average diameter is given by:

$$MeanDAP = \sqrt{(AB) \times (12732,39)} \div n$$

Where: AB is the basal area value and n is the number of individuals per hectare.

1.1.2 Basal Area of Living Trunks (AB)

The basal area of live trunks was defined as the area occupied within a plot by a live trunk with a given diameter, expressed in m^2 (square meter) per ha (hectare). The basal area serves as an indicator of the degree of development acquired by the forest, as it expresses the volume of wood and biomass in the forest.

The basal area is expressed by the following formula: $AB = \pi r^2$, where g is the basal area and r is the radius, replacing the radius with r = dap/2, then you have AB = π/4 x dap2, forming the following formula: g = π (DAP)2/ 4x10,000. To express the formula in m2 per hectare we have:

$$AB(m^2) = 0,00007854 \times (DAPcm)^2$$

To calculate the average basal area, the basal area value is divided by the number of individuals measured.

1.1.3 Tree height

The height of the trees was measured using a telescopic pole and a clinometer. The average height of the forest was obtained from the average height of all the trees counted in each plot.

1.1.4 Number of Trunks

To count the number of trunks, it was considered that when the tree bifurcates below chest height, two trunks and a single individual should be counted.

1.1.5 <u>Relative Density</u>

The relative density is the contribution in number of individuals by each forest. This measure is a comparison between the various fixed plots and is obtained by:

$$Relative\ Density = \frac{Number\ of\ individuals}{Total\ Number\ of\ Individuals} \times 100$$

1.1.6 <u>Relative Dominance</u>

Defined as the basal area of dominance between the various plots and obtained by:

$$Do\ \min\ \grave{a}ncia\ Re\ iativa = \frac{Dominance\ of\ a\ Species}{Total\ Dominance(Area\ Basai\ )} \times 100$$

1.1.7 <u>Relative Frequency</u>

Value related to the probability of occurrence of a species in a single plot. This value was calculated between plots of the same size and obtained by:

$$Reactive\ Frequency = \frac{Frequency\ of\ a\ Species}{Sum\ of\ Frequency\ of\ all\ Species} \times 100$$

1.2 Environmental degradation

On the same site where the plots are located, a checklist of environmental indicators was applied on November 3, 2009, through on-site observation, following the methodology adapted from Tommasi (1994) for estuarine areas.

This methodology consists of filling in a table where the impacts on the environment are presented in columns and their effects in rows.

Table 1. Weights, effect scores and classification of environmental impacts.

Impact weights (Pi)	Effect notes (Ne)	Rating (PixNe)
5 = extreme	-5 = extreme	-15 to -25 = extreme
3 = moderate	-3 = moderate	-5 to -9 = moderate
1 = small	-1 = small	-1 to 3 = small

Source: Tommasi (1994).

Each impact can have a weight of 1 (small), 3 (moderate) or 5 (extreme), established subjectively, according to its importance in relation to the analysis principles adopted. Impacts that had a drastic or global impact on the environment were considered extreme, while those that were significant but had more specific characteristics were considered moderate. The effects of the impacts were also given a value, but negative scores (-1, -3, -5), depending on their intensity, or a score of zero (0) when absent. The results of multiplying the weights assigned to the impacts by the scores of their effects made it possible to classify each impact into the following categories: small (values -1 to -3) moderate (values -5 to -9) and extreme (values -15 to -25) (Table 1). The sum of the values from this multiplication gives the overall impact index in the estuary studied, which is considered small (-1 to - 100), moderate (-100 to - 170) and extreme (-171 onwards).

5 RESULTS

With regard to the abiotic data analyzed, air temperatures were recorded between 24°C in plot A and 27°C in plot C, while soil temperatures remained between 26°C in plots A and B and 27°C in plot C (Table 2).

With regard to the salinity of the interstitial water, a value of 39 was obtained in plot A, while 35 was recorded in plots B and C (Table 2).

Table 2. Values corresponding to the analysis of interstitial water, air temperature and soil temperature in the different plots in the mangrove swamp of the Jaboatâo River, Pernambuco State.

Installments	Date of Collection	Tide height	Tide times	Salinity of interstitial water	Air temperature (°C)	Soil temperature (°C)	Sampling times
A	03/11/09	0,15	09h45	39	24,0	26,0	10h35
B	03/11/09	0,15	09h45	35	25,5	26,0	11h05
C	03/11/09	0,15	09h45	35	27,0	27,0	11h35

Source: Prepared by the authors (2018).

5.1 Forest Structure

The site analyzed in the mangrove swamp of the Jaboatâo River recorded the occurrence of the species *R. mangle*, *A. schaueriana* and *L. racemosa*.

Plot A (Figure 4) was made up of 1100 ind/ha with diameters ranging from 38cm to 83cm. *R. mangle* was 100% dominant (Figure 5), with an average height of 8m and a canopy height of 8.6m. The total basal area was 29.65m^2 /ha and the average was 2.69m^2 /ha, with all the individuals having a DBH $\geq$ 10.0cm and the average DBH was 61.11cm (Figure 6, Table 3).

Figure 4 - Partial view of plot **A** in the mangrove swamp of the Jaboatâo River, PE.

Source: Prepared by the authors (2018).

Figure 5: Dominance of species in each plot in the mangrove swamp of the Jaboatâo River, Pernambuco State.

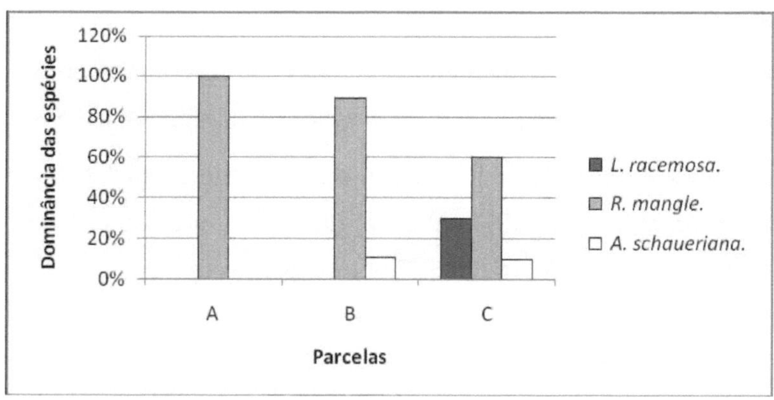

Source: Prepared by the authors (2018).

Figure 6. Average height and canopy height of the individuals in the plots analyzed in the mangrove swamp of the Jaboatào river/PE.

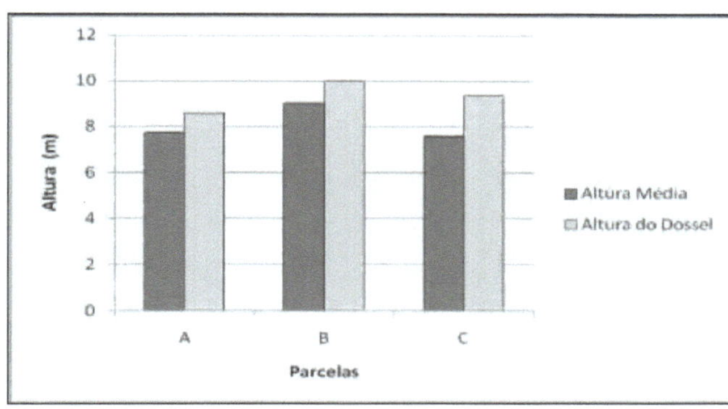

Source: Prepared by the authors (2018).

Table 3. Number of living individuals, basal area, average basal area, average DBH, average height, canopy height and average of the plots analyzed in the mangrove swamp of the Jaboatào River/PE.

PARCELS	Area number Height					Canopy height (m)
	Individuals Couple Area		DAP	Average		
	Live (m^2/ha) Basal	Average	(m)			
	(m^2/ha)	Average	(cm)			
	(m2/ha)					
A	1100	29,65	2,69	61,11	7,72	8,66
B	800	33,61	4,20	83,71	9,00	10,00
C	700	16,52	2,36	69,66	7,57	9,33
Average Plots	866	79,69	3,08	71,49	8,09	9,33

Source: Prepared by the authors (2018).

Plot B (Figure 7) totaled 800 ind/ha, with diameters ranging from 42 cm to 90 cm. Dominance was 89% *R. mangle* and 11% *A. schaueriana* (Figure 5), with an average height of 9m and a canopy height of 10m. The total basal area was 33.61m^2/ha and the average was 4.20m2/ha, with *R. mangle* having a basal area of 29.25m2/ha and A. schaueriana 4.35m2/ha. All the individuals had a DBH ≥ 10.0 cm and an average DBH of 83.71 cm (Figure 6, Table 3).

Figure 7 - Partial view of plot **B** in the mangrove swamp of the Jaboatâo River, PE.

In plot C (Figure 8), 700 ind/ha were counted, with diameters ranging from 38 cm to 80 cm. It was the plot with the greatest variety of species, with 60% of the forest made up of *R. mangle*, 30% of *L. racemosa* and 10% of *A. schaueriana* (Figure 5). The individuals had an average height of 7.57m and a canopy height of 9.33m, the total basal area was 16.52m^2/ha and the average was 2.36m^2/ha (Figure 6, Table 3). Considering the basal area per species, *R. mangle* had 12.27m2/ha, *L. racemosa* 2.41m2/ha, and *A. schaueriana* 1.83m2/ha. Of the ten individuals quantified in the plot, two had DBHs in the ≥2.5cm and <10.0cm class and the others, ≥ 10.0 cm, with an average DBH of 69.66cm.

Figure 8. Partial view of plot **C** in the mangrove swamp of the Jaboatâo River/PE.

With regard to the relative density of the species in the three plots analyzed, *R. mangle* accounted for 84%, *L. racemosa*, 10%, while *A. schaueriana* contributed 6% (Figure 9).

In terms of relative dominance, *R. mangle* accounted for 89%, while *A. schaueriana* and *L. racemosa* accounted for 8% and 3%, respectively (Figure 10).

In terms of relative frequency, *R. mangle* showed 50%, *A. schaueriana*, 33% and *L. racemosa*, 17% (Figure 11).

Figure 9. Relative density of the species in the plots analyzed in the mangrove swamp of the Jaboatâo River/PE.

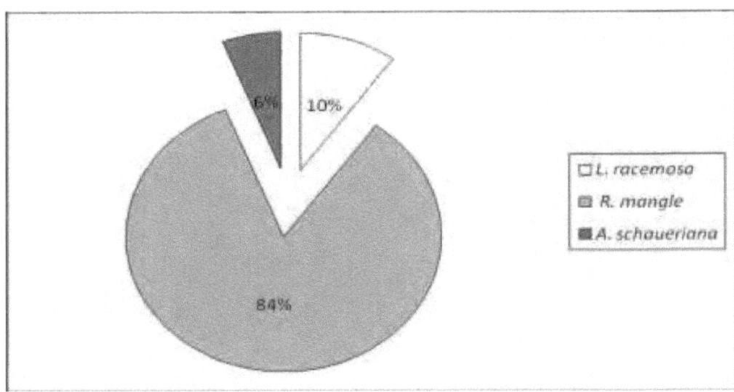

Source: Prepared by the authors (2018).

Figure 10. Relative dominance of the species in the plots analyzed in the mangrove swamp of the Jaboatâo River/PE.

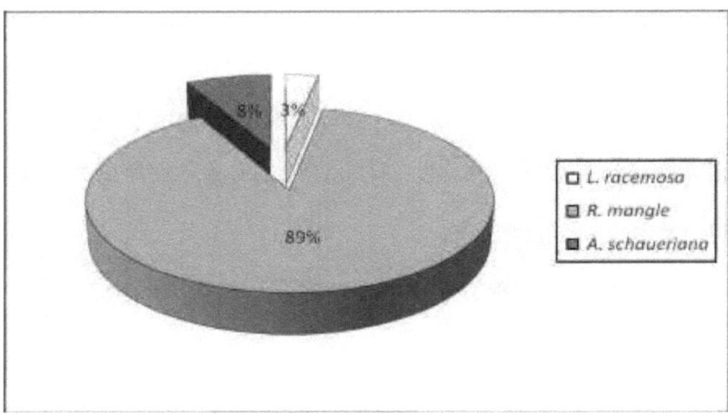

Source: Prepared by the authors (2018).

Figure 11. Relative frequency of species in the plots analyzed in the mangrove swamp of the Jaboatâo River, Pernambuco State.

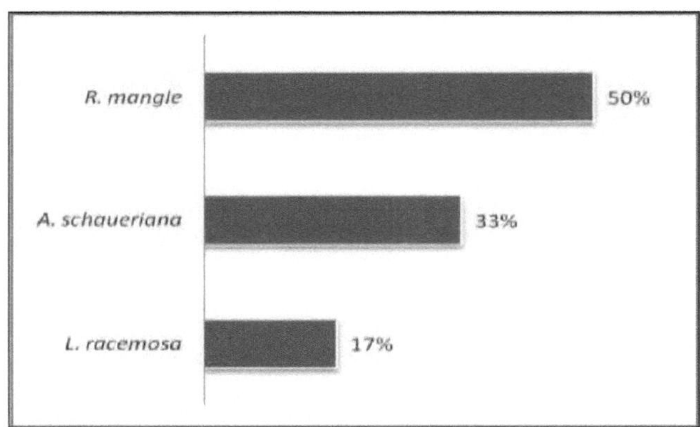

Analyzing the floristic composition of the three plots studied, it was possible to observe that only specimens of *R. mangle* were found in plot A, in B this last species and *A. schaueriana* occurred, while in C the presence of *L. racemosa* was recorded in addition to these two, defining a zonation pattern in the forest (Figures 5 and 12).

A structural comparison between the plots showed that plot A had the highest number of individuals and the lowest canopy height (Table 3).

Plot B showed greater structural development as it had the highest values for all the structural parameters assessed (Table 3).

In plot C, on the other hand, the lowest structural values were recorded, with the exception of average DBH and canopy height (Table 3), but the highest density of dead individuals was found (Figure 13).

Figure 12. Basal area of species in the mangrove forest of the Jaboatâo river/PE.

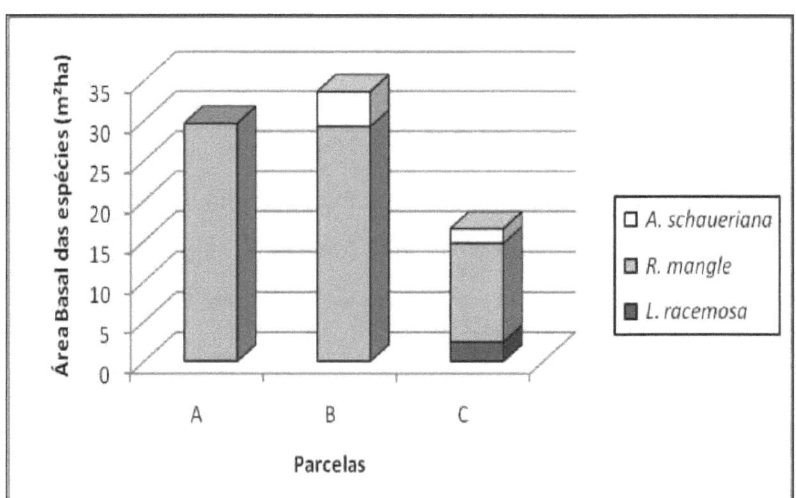

Figure 13. Density of living and dead individuals in the mangrove swamp of the Jaboatâo River, PE.

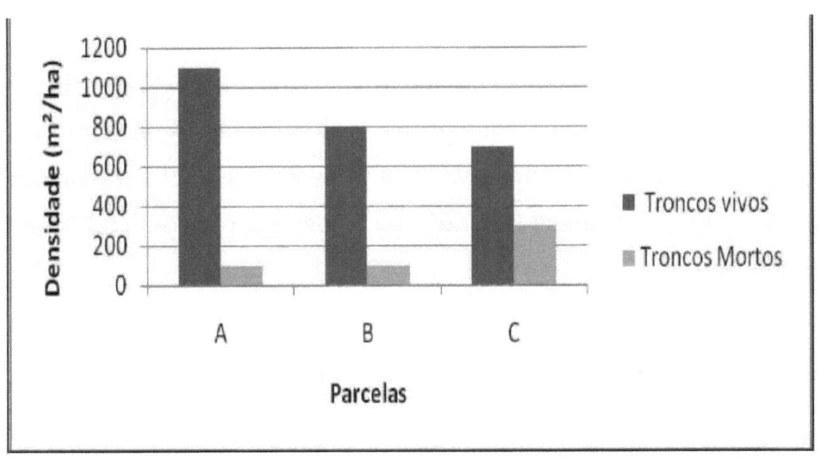

Source: Prepared by the authors (2018).

5.2 Environmental degradation

According to the data obtained by applying the *checklist*, the indicator that was considered to have the greatest environmental impact, with a class number of -25, was the emission of domestic and industrial effluents. Those classified as moderate, with values between -5 and -9, were: urban expansion, landfill in mangroves, burning of adjacent vegetation, logging in adjacent forests, recreation and ports/marinas.

Thus, by adding up the resulting values for each indicator, the impact index for the stretch analyzed was considered small, with a value of - 90 (Table 4).

Table 4. *Checklist* of anthropogenic pressure indicators applied to the Ponta Cabo portion of the Jaboatâo River mangrove swamp.

INDICATORS	WEIGHT	EFFECT	CLASS
Urban sprawl	3	-3	-9
Earthworks	1	Absent	Absent
Landfills in mangroves	3	-3	-9
Roads and highways in the mangroves	1	Absent	Absent
Mangrove paths	1	-1	-1
Erosion processes	1	Absent	Absent
Waste disposal	1	-1	-1

Extensive farming	1	Absent	Absent
Subsistence farming	1	Absent	Absent
Vegetation degradation	1	-1	-1
Burning of vegetation adjacent to mangroves	3	-3	-9
Burning mangroves	1	-1	-1
Logging in the forests adjacent to the mangroves	3	-3	-9
Logging in the mangroves	1	-1	-1
Death of the mangrove	1	-1	-1
Domestic and industrial effluent emissions (estuary/marsh)	5	-5	-25
Recreation (estuary/marsh)	3	-3	-9
Artisanal fishing (estuary/marsh)	1	-1	-1
Siltation (river/marsh)	3	-1	-3
Bridges	1	Absent	Absent
Invasion of public areas (river/marsh)	1	-1	-1
Irrigation (river)	1	Absent	Absent
Ports/marinas	3	-3	-9
Shrimp nurseries	1	Absent	Absent
Dams and/or channel obstructions	1	Absent	Absent
Canalization of the riverbed	1	Absent	Absent
Artificial opening of the estuary	1	Absent	Absent
RESULT OF THE SUM			**-90**

Source: Prepared by the authors (2018).

6 DISCUSSION

The structure of mangrove forests is the result of the environmental factors to which they are subjected; many forces of different intensities and frequencies act in these environments (SCHAEFFER- NOVELLI, 2002).

With regard to temperature, the values recorded in the Jaboatao river mangrove remained above 24°C, which is in line with the ideal conditions for mangrove development, i.e. average temperatures above 20°C. In these ecosystems, the annual temperature range should be less than 5°C, with average minimum temperatures of no less than 15°C (KIENER, 1973; DIEGUES, 1987; SCHAEFFER-NOVELLI, 1995; VANNUCCI, 2002).

The salinity values of the interstitial water were high in plot A (39), while in plots B and C they remained at 35.

Barbosa (2010) recorded salinity values in the Pina/PE mangrove that only exceeded 0.5 in one of the nine plots analyzed, which may be explained by the location of his study sites, situated in areas with less marine influence.

Nascimento Filho (2007), in his study of the Ariquindà River, Tamandaré/PE, found salinity values ranging from 40 to 90. The author emphasizes that salinity, associated with topography, was considered the best parameter for defining zonation patterns, since it can be easily measured and correlated directly with the distribution of species.

In the estuary of the river Sâo Mateus/ES, Silva; Bernini and Carmo (2005) found salinity values that varied between 2 and 38.

Fernandes and Peria (1995) point out that mangrove species are typical of saline environments, although they can grow in environments free from the presence of salt, under these conditions the formation of forests does not occur, as the mangrove species end up losing space to those that grow faster and are better adapted to the presence of fresh water. According to these authors, *Rhizophora* is the least salt-tolerant genus, growing in environments where the interstitial water content is lower than 50. *Avicennia is* the most tolerant genus, managing to survive in places where the interstitial water can vary from 65 to 90 and *Laguncularia* shows intermediate tolerance when compared to the two previous genera.

6.1 Forest Structure

Three species of mangrove were identified for the mangrove swamp of the Jaboatâo River in its portion known as Ponta Cabo: R. mangle, L. racemosa and A. schaueriana.

According to Coelho et al. (2004), the Pernambuco coast is home to the three typical mangrove genera (Rhizophora, Avicennia and Laguncularia) and the genus considered to be transitional (Conocarpus).

In this way, the floristic composition currently found corroborates the results obtained by other studies in Pernambuco mangroves, such as Schuler; Andrade and Santos (2000), in the Ariquindà River estuary/PE, by Correia (2002), in the Santa Cruz Canal mangrove, by Nascimento Filho (2007), in the Timbó River estuary and Barbosa (2010), in the Pina mangrove/PE. In relation to Suape, Souza and Sampaio (2001) also refer to the occurrence of A. germinans.

The species showed a defined zonation pattern; R. mangle appeared predominantly on the mangrove fringe, with the species A. schaueriana and L. racemosa gradually integrating into the forest further inland.

According to Coelho et al. (2004), the distribution pattern is not always regular, but the red mangrove would be more common in the part closest to the sea, the button mangrove in the outer margin of the mangrove, the siriùba mangrove in the middle portion and the white mangrove in the portion furthest from the sea, upstream. This distribution may have been modified many times, either by natural events or human intervention.

Undefined zonation patterns were found in Suape/PE (SOUZA; SAMPAIO,

2001), Ariquindâ/PE (NASCIMENTO-FILHO, 2007) and Pina/PE (BARBOSA, 2010), which, according to Cintrón and Schaeffer-Novelli (1983), is common in mangroves in the Northeast.

The relative density in the area studied showed that R. mangle had the highest indices, followed by L. racemosa and, with a lower density, A. schaueriana. The same fact can be observed in the mangrove swamp of the Timbó River/PE (CORREIA, 2002), where R. mangle stood out as the predominant species, followed by L. racemosa and, to a lesser extent, A. schaueriana. In the Pina/PE mangrove swamp, according to Barbosa (2010), L. racemosa showed the highest density indices, followed by R. mangle and then A. schaueriana.

Deus et al. (2003) found different species and densities in mangroves in Piaui, where the highest density was L. racemosa, followed by A. germinans and, lastly, R. mangle.

In the mangrove swamp of Peninsula Bragança/PA (MENESES; MEHLIG, 2006), the species R. mangle, A germinans and L. racemosa occurred, with R. mangle being the most abundant, followed by A. germinans and, to a lesser extent, L. racemosa.

In relation to the total density of living individuals in each plot in the area studied, the values reached 1,100 ind/ha in plot A. Plot C, despite having the lowest number of live individuals, had the highest number of dead individuals, both of the L. racemosa and R. mangle species, which may be the result of competition between the individuals. It is worth noting that this was the only plot where all three species found on the site occurred.

Silva; Bernini and Carmo (2005) found densities ranging from 450 to 1,450 ind/ha in the estuary of the Sâo Mateus river, and this increasing variation was biased towards the river and the sea.

While in 32 forests, distributed over six transects, in Guanabara Bay, Soares et al. (2003) found species with total densities ranging from zero (clearings) to

52,800 individuals/ha.

In the Pina/PE mangrove swamp, Barbosa (2010) found density values of up to 2,850 ind/ha, but a high number of cuts were found in the low-density plots.

The density of a forest is a function of its age and maturity, i.e. during the course of its development, forests pass from a phase in which the land is occupied by a high density of small trees, to a more mature phase, when a few large trees dominate, i.e. the density decreases as the forest matures. The process that causes this reduction is due to competition between the canopies for space, as well as root development. The tallest trees receive direct sunlight and grow rapidly, hindering or even preventing the development of those individuals whose canopies do not receive direct light, who die due to competition, leaving even more space for the development of the better endowed ones (SCHAEFFER- NOVELLI; CINTRÓN, 1986).

The total basal area of the site studied in the Jaboatâo River Mangrove was 79.78 m²/ha, where the dominant species was R. mangle, which was present in all the plots analyzed, followed by A. schaueriana and, to a lesser extent, L. racemosa.

The highest biomass and wood value of the Jaboatâo river mangrove was 33.61m2/ha of basal area, in plot B, which had the lowest number of individuals. This indicator is an excellent gauge of the degree of development of the forest, since it is closely related to the volume of wood and the biomass of the forest (SCHAEFFER-NOVELLI; CINTRÓN, 1986).

Bernini and Rezende (2004), studying the Paraiba do Sul River/RJ *mangrove* swamp, found that the total basal area of living individuals was 40m²/ha, with A. germinans being the dominant species, with 60%, followed by *R. mangle* (25%) and *L. racemosa*, with 15%. Silva; Bernini and Carmo (2005) in the estuary of the São Mateus River in Espírito Santo found different dominance in their study sites. Those under greater influence of the tides showed less structural development and dominance of the *R. mangle* species, while those with better structural development were dominated by L. racemosa and *A. germinans*. In the Pina/PE mangrove swamp (BARBOSA, 2010), lower basal area values were found, reaching a maximum of 30.86m2/ha, while in the Jaboatao river mangrove swamp, a higher maximum value was recorded. This difference in basal area may be related to the greater contribution of R. mangle in the Jaboatao river mangrove, since the dominant species in the Pina/PE mangrove is L. racemosa.

Frequency is the percentage of plots in which a particular species is found, but comparisons between plots can only be made between plots of the same size (SCHAEFFER-NOVELLI, 1986).

In the mangrove swamp of the Jaboatao River, the most frequent species was *R. mangle*, followed by *A. shaueriana* and the least frequent *L. racemosa*.

Souza and Sampaio (2001) in the mangroves of Suape/PE stated that the most frequent species was *R. mangle*, present mainly on the mangrove fringe, followed by *A. schaueriana* and/or *L. racemosa*.

The average values for average height and canopy height were 8.1m/9.3m respectively, with plot B having the highest average height and canopy height, while the maximum height remained the same at 10m respectively.

In the Piaui mangrove swamp, Deus et al. (2003) found maximum heights ranging from 12 to 28m and an average height of 11.2m, with most individuals having heights below 2m or around 10m. Bernini and Rezende (2004), studying the mangrove swamp of the Paraiba do Sul River in Rio de Janeiro, observed that the height of the forest varied from 1.10 to 19.5m, with an average of 3.9m to 8.9m. Nascimento Filho (2007), studying the Ariquindà river mangrove (PE), found average forest height ranging from 0.4 to 8.9m and canopy height from 1.4 to 11.1m. In relation to the Pina/PE mangrove swamp (BARBOSA, 2010), he found average heights varying between 6.3m and 9.4cm and canopy heights between 8.0m and 14.5m.

The above shows that the site studied in the mangrove swamp of the Jaboatâo River has the characteristics of a developed forest, since it is dominated by a few trees with a considerable biomass value, as well as having a defined zonation pattern.

6.2 Environmental degradation

According to Noriega (2004), the Jaboatâo River estuary is an important water body for the development of diverse activities such as recreation, leisure areas and a natural source for the cultivation of aquatic species. However, this estuary has suffered over the years from the pressure of urban and industrial development, representing one of the areas most vulnerable to degradation caused by increased urban and real estate pressure in the state of Pernambuco.

The survey of anthropogenic pressure points in the stretch of mangrove analyzed in the Jaboatâo River estuary, however, made it possible to classify this stretch as having little impact (-90). However, the discharge of domestic and industrial effluents was considered extreme in the *checklist* classification.

Leite (2009), carrying out an environmental and fisheries survey of the Jaboatâo and Pirapama rivers, found that as well as industrial pollution of the waterways, there is also domestic pollution, both from sewage and from garbage dumps on the river banks. According to this author, the chemical contaminants in the effluents discharged by industries and, above all, urban sewage with a high organic matter content, are not only diverse but also abundant, representing a great polluting pressure for the body of water and the aquatic organisms that live in it.

In this checklist, urban expansion was considered to have little impact, given that the area under analysis refers to the right bank of the Jaboatâo River, and it has only one house which, according to the owner, has permission from IBAMA to live inside the mangrove (Figure 14). However, this scenario is completely different on the left bank, where, according to Silva (2009), the evolution of urban expansion was quite significant between 1974 and 1997.

Figure 14, Household inside the mangrove swamp of the Jaboatâo River, PE.

Source: Prepared by the authors (2018).

Bryon (1994) states that in 1991 the urbanization rate of the RMR reached 95%, a slight reduction when compared to the 1980 rate of 96%. Despite the decrease, Jaboatâo dos Guararapes had a high demographic density.

The other indicators of anthropogenic pressure listed in the *checklist* were considered moderate or not very significant.

Some authors (BRYON, 1994; NORIEGA, 2004; SILVA; OLIVEIRA; TORRES 2009) highlight waste disposal, extensive and subsistence agriculture, irrigation, among others, as significant drivers in the estuary studied.

Recent studies carried out in mangroves in Pernambuco have assessed the level of degradation of these areas using a *checklist*.

Melo (2010), studying the mangroves of the lower course of the Capibaribe River in the state of Pernambuco, identified the level of environmental degradation as extreme in most of the sectors analyzed.

Santana (2010), researching the mangrove swamp of the Itapessoca River in the state of Pernambuco, found that the estuary is an extremely impacted environment.

However, it should be emphasized that this study only covered the right bank of the Jaboatâo River, in the area known as Ponta Cabo, where real estate expansion has so far been negligible, which has contributed to mangrove conservation.

This study can be considered of great relevance to the knowledge of the mangroves distributed in the Jaboatâo river estuary, because although it is a preliminary survey, it is unprecedented in terms of the composition and structure of the mangrove forest.

It is also important to point out that the data obtained here could serve as a subsidy for management and monitoring actions in the area, since it is close to large-scale constructions, such as the Praia do Paiva Tourist Complex, which consists of the implementation of an international destination associated with high-standard residential, commercial and service developments, located in the municipality of Cabo de Santo Agostinho (Figure 15).

Figure 15. Paiva Beach Tourism and Leisure Complex. Source: Pernambuco State Government.

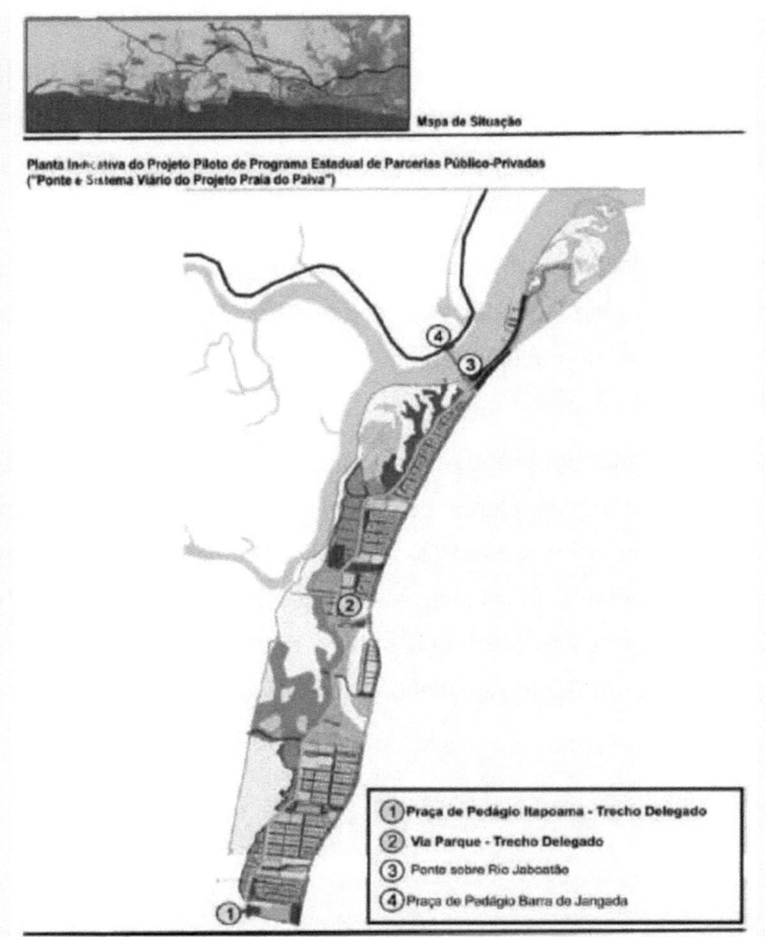

Source: Prepared by the authors (2018).

7 CONCLUSIONS

The abiotic data recorded at the site studied in the Jaboatao river mangrove fall within the parameters considered characteristic for the mangrove ecosystem.

The presence of the species *R. mangle*, *L. racemosa* and *A. schaueriana* in the mangrove swamp analyzed corroborates the results obtained by various studies carried out in estuarine areas of the Brazilian Northeast.

The species showed a defined zonation pattern, with *R. mangle* predominating in the fringe of the forest, gradually being integrated by the species *A. schaueriana* and, later, *L. racemosa* further into the mangrove, which is not always observed in other mangroves that have already been studied.

The structural parameters of the forest indicate that it is in a mature phase, as it is composed of few trees, but of great size and volume, with the *Rhizophora mangle* species predominating.

The site studied can be considered conserved, with a low level of degradation, due to its location on the right bank of the Jaboatao River, where anthropogenic factors are still incipient.

Finally, this study can be considered of great relevance to the knowledge of the mangroves distributed in the Jaboatao river estuary, because although it is a preliminary survey, it is unprecedented in terms of the composition and structure of the mangrove forest, making it essential for future management and monitoring actions.

REFERENCES

AYOADE, J.O. **Introduction to climatology for the tropics.** Coordinated by Antonio Christofoletti. 11.ed. Rio de Janeiro: Bertrand Brasil, 2006.

BARBOSA, F. G. **Structure and space-time analysis of the vegetation of the Pina mangrove swamp, Recife-PE:** subsidies for management, monitoring and conservation. Dissertation - Federal University of Pernambuco. CFCH. Geography. 89 f.: ill., fig., tab. Recife: 2010.

BARRETO, D. S. P.; CANTARELLI, J. R. R.; CIRILO, J. A.; LIMA, J. C. A. L. Characterization of the use and occupation of the Jaboatâo river estuary/PE-Brazil using high resolution images. In: **Inter-American Congress of Sanitary and Environmental Engineering**, Punta Del Este, 2006.

BRANCO, E.; FEITOSA, F. A. N.; FLORES MONTE, M. J. Seasonal and spatial variation of phytoplankton biomass related to hydrological parameters in the Barra das Jangadas estuary (Jaboatao dos Guararapes - Pernambuco - Brazil). **Tropical Oceanography** (Revista Online), Recife, v. 30, n. 2. p. 79-96, 2002.

BRYON, M. E. Q. **Desenvolvimento Urbano x Meio Ambiente:** a relação da ocupaçao do espaço urbano com os recursos naturais remanescentes o caso das áreas estuarinas da regiao metropolitana do Recife (RMR).Dissertaçao de Mestrado, Programa de Pós-graduaçao em Desenvolvimento Urbano e Regional, CAC - UFPE, 1994.

CINTRÓN, G; SCHAEFFER-NOVELLI, Y. **Proposal for the study of marsh and mangrove resources.** Internal Report of the Oceanographic Institute of the University of Sao Paulo, No. 10. Sao Paulo, 1981. p. 1-13.

CITRON, G. SCHAEFFER-NOVELLI, Y. **Introdución a la ecologia del manglar.** Montevideo, UNESCO/ ROSTLAC. 109p. 1983.

COELHO, P. A. Distribution of decapod crustaceans in the Barra das Jangadas area. **Trab. do Instit. Oceanogr. Univ. Fed. de Pernambuco**. Recife: v. 5 - 6, 1963/64. p. 159 - 174.

COELHO, P.A; BATISTA-LEITE. L. M. A; SANTOS,M. A. C; TORRES, M. F. A. The Mangrove. *In :* ESKINAZI-LEÇA, E.; NEUMANN-LEITÂO, S.; COSTA, M. F.(Orgs). **Oceanography in a Tropical Setting**. Bagaço: Recife, 2004. 761 p.

CORREIA, V.L. **O Bosque de mangue do estuàrio do rio Timbó, estado de Pernambuco, Brasil:** Caracteristicas estruturais e Vulnerabilidade da área frente às tensões antrópicas. Master's dissertation, Regional Postgraduate Program in Development and Environment/PRODEMA, 2002.

COUTINHO, P. N. (Coord.) **Study of marine erosion on the beaches of Piedade, Candeias and the Barra de Jangadas estuary.** LGGM/UFPE, Final Report, Recife, 1997.

CPRH (PERNAMBUCO WATER RESOURCES COMPANY). **Final water resources diagnostic report (RDRH) for GL Basin 2.** Pernambuco State Government. 2006.

CUNHA, A.; VICTOR-CASTRO, F.; LIRA, L.; LARRAZABAL, M. E. L. FONSECA-GENEVOIS, V. Morphodynamics of the mouth of the Jaboatao River estuary and adjacent beaches - PE. **VII COLACMAR.** v.1, p.218-219, 1997.

DEUS, M. S. M.; SAMPAIO, E. V. S. B.; RODRIGUES, S. M. C. B.; ANDRADE, V. C. Structure of woody vegetation in three mangrove areas of Piaui with different histories of anthropization. **Brasil Florestal**, n. 78, 2003. p. 53-60.

DIEGUES, A. C. S. **Conservation and sustainable development of coastal ecosystems in Brazil.** Secretariat of the Environment of the State of São Paulo, 1987.

EMBRAPA (Brazilian Agricultural Research Corporation). **Sistema brasileiro de classificaçâo dos solos**, 2ª ed. - Rio de Janeiro: Embrapa Solo, 2006.

ENGESAT. **Satellite image map:** environmental studies of the metropolitan center of the coastal zone of Pernambuco. ASTER sensor, bands 1,2,3. Scale 1:25,000. Recife, October 2005.

FERNANDES, A.J; PERIA, L.C.S. In: **Mangrove:** ecosystem between land and sea. Caribbean Ecological Research: Sao Paulo, 1995. 64 p.

GERCO. **Diagnosis of the physical and biological environment and map of land use and occupation of the metropolitan center of the Pernambuco coast.** SECTMA; CPRH: 2006.

HAZIN. F. H.V.; NETO. F. F. P.; LEITE. A. P. A.; **Environmental diagnosis of the Pirapama and Jaboatao rivers.** Federal Rural University of Pernambuco. Department of Fisheries and Aquaculture. Fisheries Technology Laboratory. Recife, 2008.

KIENER, A. Les mangroves duglobe: Aspects écologiques, biocénotiques et physiologiques particuliers, mise em valeur. **Bulletin de Muséum National D'Histoire Naturelle.** 3ª series, n° 164, mai-jun 1973.

LEITE, A. P. A. **Environmental and fisheries survey of the Jaboatao and Pirapama rivers in the state of Pernambuco, Brazil.** Dissertation - Federal Rural University of Pernambuco. Department of Fisheries and Aquaculture. 67 f.: il. Recife: 2009.

MATNI, A. S.; MENEZES, M. P. M.; MEHLIG, U. Structure of the mangrove forests of the Bragança peninsula, Parà, Brazil. **Boletim do Museu Paraense Emilio Goeldi**, Ciências Naturais, Belém, v. 1, n. 3, 2006. p. 43-52.

MELO, J. G. da S. **Mangroves of the lower Capibaribe River:** vegetation structure, environmental impacts and space-time analysis. Monograph - Federal University of Pernambuco. CFCH. Geography. 70 f.: ill., fig., tab. Recife: 2010.

MONTEIRO, L. H. U.; SOUZA, G. M.e;MAIA, L. P.; LACERDA, L. D. de.; Evolution of mangrove areas on the northeast coast of Brazil between 1978 and 2004.**Revista da Associaçao Brasileira de Criadores de Camarao**, Recife, p. 36 - 42 01 set. 2004.

MOREIRA, H. A. **Diagnosis of the environmental quality of the Jaboatao River Basin: suggestion for a preliminary framework.** Dissertation - Federal University of Pernambuco. DCG. Postgraduate Program in Civil Engineering, 2007.

NASCIMENTO FILHO, G. A. **Structural development and zonation pattern of mangrove forests in the Ariquindà River, Tamandaré Bay, Pernambuco, Brazil.** Recife, 2007. 82 p.

NORIEGA, D. C. E. **Hydrological influence and degree of pollution of the Pirapama and Jaboatao rivers in the Barra das Jangadas estuary (PE-Brazil):** temporal cycle. 2004.162 f. Dissertation (Master's Degree in Oceanography) Center for Technology and Geosciences. Federal University of Pernambuco, Recife, 2004.

NORIEGA, C. D. E. **Spatial distribution of phytoplankton biomass and its relationship with nutrient salts in the Barra das Jangadas estuarine system (Pernambuco - Brazil)** Arq. Ciên. Mar, Fortaleza, 2005.

NORIEGA, C. D. E; MUNIZ, K.; ARAÙJO, M. C.; TRAVASSOS, R. K.; NEUMANN- LEITÂO, S. Fluxes of dissolved inorganic nutrients in a tropical estuary - Barra das Jangadas - PE, Brazil.**Tropical Oceanography** (Revista Online), Recife, v. 33, n. 2. p. 129-139, 2005.

OKUDA, T.; NÓBREGA, R. Study of Barra das Jangadas. Part. I - Distribution and Movement of Chlorinity - Quantity of Current. **Paper from the Institute of Marine Biology and Oceanogr. University of Recife.** Recife: v. 1. n. 2. p. 175-191, 1960.

OKUDA, T.; CAVALCANTI, L. B.; BORBA, M. P. Study of Barra das Jangadas Part. II - Variation in pH, dissolved oxygen and permanganate consumption. **Paper from the Institute of Marine Biology and Oceanogr. University of Recife.** Recife: v. 1. n. 2. 60p. 193, 1960.

OKUDA, T.; CAVALCANTI, L. B.; BORBA, M. P. Estudo da Barra das Jangadas Part. III - Variation of nitrogen and phosphate.**Trab. do Instit. de Biologia Maritima e Oceanogr. University of Recife.** Recife: v. 1. n. 2. p. 207, 1960.

OTTMANN, F.; OTTMANN, J. M. Study of Barra das Jangadas Part. IV- Study of the sediments. **Work of the Institute of Maritime Biology and Oceanogr. University of Recife.** Recife: v. 1. n. 2. p. 219, 1960.

OTTMANN, F.; OKUDA, T.; CAVALCANTI, L. B.; SILVA, O. C.; ARAÙJO, J. V. A.; COELHO, P. A.; PARANAGUA, M. N.; ESKINAZI, E. Study of Barra das Jangadas - Part V - Effects of pollution on the ecology of the estuary.**Trab. do Instit. Oceanogr. Univ. Fed. de Pernambuco.** Recife: v. 7 - 8. p. 7 16, 1965/66.

PROST, M. T.; RABELO, B. V. Phytospatial variability of coastal mangroves and coastal dynamics: examples from French Guiana, Amapà and Parâ. **Bol. Mus. Para. Emilio Goeldi,** ser. Ciência da Terra, v. 8, p. 101-121, 1996.

SANTANA, N. M. G. **Mangroves of the Itapessoca-Goiana-PE river estuary:** Space-Time Analysis and Environmental Degradation. Monograph - Federal University of Pernambuco. CFCH. Geography. 57 f.: ill., fig., tab. Recife: 2010.

SCHAEFFER-NOVELLI, Y. (Coord). **Mangroves**: ecosystem between land and sea. Caribbean Ecological Research: Sâo Paulo, 1995. 64 p.

SCHAEFFER-NOVELLI, Y. **Mangrove: an** ecosystem that goes beyond its own borders. In: Congresso Nacional de Botânica, 53, Recife, p. 34-37, 2002.

SCHAEFFER-NOVELLI, Y; CINTRÓN, G. **Guide to mangrove area studies (structure, function and flora).** [S.I.]: Caribean Ecological Research. 1986. 156 p.

SCHULER, C. A. B.; ANDRADE, V. C.; SANTOS, D. S. The mangrove: composition and structure. In: BARROS, H. M.; ESKINAZI-LEÇA, E; MACEDO, S. J; LIMA, T. **Gerenciamento participativo de estuârios e manguezais.** Recife: Ed. Universitària da UFPE, 2000 p.27-38

SILVA, M. C. **Relationships between coastal dynamics and meiofauna in an impacted environment (Jaboatao River Estuary, Pernambuco, Brazil).** Master's dissertation, presented to the Zoology Department of UFP. Recife, 1997.

SILVA, O. C.; COELHO, P. A. Estudo ecológico da Barra das Jangadas (Nota Prèvia).**Trab. do Instit. de Biologia Maritima e Oceanogr. Univer. do Recife.** Recife: v. 1. n. 2. p. 235, 1960.

SILVA, M. A. B; BERNINI, E.; CARMO, T.M.S. Structural characteristics of mangrove forests of the Sao Mateus river estuary, ES, Brazil. **Acta Botànica Brasilica**.19(3), 2005. p. 465-471.

SILVA, J. S.; OLIVEIRA, T. H.; TORRES, M. F. A.. **Use of aerial images in the temporal analysis of impacted environments: case study - Jaboata river estuary - Pernambuco - Brazil**. In: XIV SBSR, 2009, Natal- RN. XIV SBSR Remote Sensing. Sao José dos Campos : INPE, 2009. v. 1. p. 6257-6283.

SOARES, M. L. G.Vegetation structure and degree of disturbance in the mangroves of Lagoa da Tijuca, Rio de Janeiro, RJ, Brazil.**Revista Brasileira de Biologia**, v. 59, n. 3, 1999. p. 503-515.

SOARES, M. L. G.; CHAVES, F. O. CORRÊA, F. M.; JÚNIOR, C. M. G. S. **Structural diversity of mangrove forests and its relationship with disturbances of anthropogenic origin: the case of Guanabara Bay (Rio de Janeiro)**. Anuàrio do Instituto de Geociências-UFRJ, v. 26, 2003.p. 101-116.

SOUZA, M. M. A.; SAMPAIO, E. V. Physiographic types of the mangroves of Suape-PE-Brazil: vegetation and sediment structure. **In: Mangrove 2000** - Sustainability of estuaries and mangroves: challenges and perspectives. Full papers. Recife, CD-ROM, 2000. p.1-11.

TOMMASI, L. R. **Estudo de impacto ambiental.** Sao Paulo: CETESB, Terragraph Artes e Informàtica, 1994. 354 p.

VANNUCCI, M. **Os Manguezais e nós**: uma sintese de percepções. 2ª ed. rev. e amp. Sao Paulo: EDUSP, 2002. 244 p.

WALSH, G. E. 1974. Mangroves: a review. *In*: REIMOLD, R.J and QUEEN, W.H. (Eds.). **Ecology of halophytes.** Academic Press, New York, p. 51-171.

(ZAPE) **Agroecological Zoning of the State of Pernambuco.** Recife: Embrapa Solos - research and development execution unit - UEP Recife; Pernambuco State Government, 2001. CD-ROM (Embrapa Solos. Documentos, 35).

FOREST CODE. Available at

<http://www.controleambiental.com.br/código_florestal.htm> Accessed: May 2010.

GOVERNMENT OF THE STATE OF PERNAMBUCO. **Bridge and road system for the Praia do Paiva project.** . Available at <http://www.ppp.seplag.pe.gov.br/web/ppp/projetos-ppp> Accessed: June 2010.

FEDERAL UNIVERSITY OF CAMPINA GRANDE. **Department of Atmospheric Sciences.**

Rainfall data. Available at < http://www.dca.ufcg.edu.br/ >Accessed on: May 2010.

I want morebooks!

Buy your books fast and straightforward online - at one of world's fastest growing online book stores! Environmentally sound due to Print-on-Demand technologies.

Buy your books online at
www.morebooks.shop

Kaufen Sie Ihre Bücher schnell und unkompliziert online – auf einer der am schnellsten wachsenden Buchhandelsplattformen weltweit! Dank Print-On-Demand umwelt- und ressourcenschonend produziert.

Bücher schneller online kaufen
www.morebooks.shop

Printed by Books on Demand GmbH, Norderstedt / Germany